COMComponentConstruction

AdvancedTechniquesForEfficientSoftwareIntegration

Evelyn Mcknight

Table of Contents

1 Introduction to COM

COM, at its core, is about creating reusable software components. Think of it like building LEGO bricks: you create individual, self-contained pieces that can be combined in countless ways to construct larger, more complex structures. These "bricks" are your COM components, and they interact through precisely defined interfaces, ensuring seamless integration regardless of their internal workings. Understanding this fundamental concept is the key to mastering COM. This modularity promotes code reusability, simplifying development and maintenance considerably, leading to more efficient software projects overall. The power of COM lies in its ability to transcend language and platform barriers, fostering interoperability that's truly remarkable in the software development world.

The beauty of COM lies in its interface-based architecture. Imagine you're ordering a pizza. You don't care how the pizza is made, just that it arrives as ordered. Similarly, a COM component exposes an interface—a contract—specifying what it can do, not how it does it. Clients interact solely through this interface, abstracting away the component's internal implementation details. This is crucial for flexibility and maintainability. Changing a component's internal workings won't affect clients as long as the interface remains consistent. This approach drastically reduces dependencies, allowing for easier upgrades and modifications throughout the lifespan

of a software project. Changes in the internal implementation will not affect the client code that utilizes the component.
Defining these interfaces is done using the Interface Definition Language (IDL). IDL is a formal language that describes the component's functionality in a platform-independent way. This allows COM components built on different platforms (Windows, for example) to communicate effectively. It's a standardized language that acts as a bridge between various programming languages and operating systems, promoting the development of components that transcend specific environments. Through IDL, you declare the functions, methods, and properties that the component will offer. This clear specification is crucial for ensuring interoperability and preventing unexpected behaviors, which is especially important in larger, more complex projects.
Consider a scenario where you need to integrate a weather reporting service into your application. Using COM, you can create a weather component with an interface that provides functions to retrieve current temperature, humidity, and forecast data. Your application, written in any COM-compatible language, can then use this component without knowing anything about its inner workings, provided that the interface specifications are met. This illustrates the power of abstraction, simplifying the development process and enhancing the system's flexibility.
The importance of proper interface design cannot be overstated. A well-designed interface is clear, concise, and focused on a specific set of functionalities. Avoid creating overly complex interfaces. Think about the future needs of the project and consider extensibility. Overly broad interfaces can lead to tightly coupled components, making them difficult to maintain and update. A modular design ensures better integration and minimizes complexities. This leads to a more maintainable codebase.

Beyond the basics, understanding the nuances of aggregation and containment is key to creating efficient and reusable components. Aggregation allows you to combine multiple components into a single, more powerful unit, while containment simply means one component uses another as a member. Understanding these relationships is essential for creating complex systems, enabling efficient code reuse and promoting maintainability. Strategic use of these techniques improves your design and reduces overall complexity.
Finally, robust error handling is paramount. Your COM components need to anticipate and gracefully handle errors, preventing application crashes and providing useful information to help with debugging and troubleshooting. It is vital for application stability and provides valuable debugging information. Implement error handling at every stage of component development to ensure that your applications can deal with unexpected situations. A well-designed error handling mechanism makes the application more robust and reliable.

1.1 COM Fundamentals

The Component Object Model, or COM, is a binary standard defining how software components interact. Understanding its fundamental principles is crucial for building robust and efficient applications. At its heart, COM relies on interfaces—clearly defined contracts specifying the methods a component exposes. These interfaces ensure loose coupling, allowing components to evolve independently without breaking existing applications. This is achieved through a system of globally unique identifiers (GUIDs), which uniquely identify each interface and component, preventing naming conflicts and ensuring interoperability across diverse environments. The implications of this architecture ripple across countless applications and systems; it is the bedrock upon which many successful software solutions have been built. Successfully leveraging COM requires understanding this fundamental structure and its underlying principles.

COM's architecture promotes modularity and reusability. Imagine building a house; you wouldn't create every brick from scratch. Instead, you use pre-made components—bricks, windows, doors—that are assembled into a functional whole. COM facilitates this same approach in software development. Individual components, each responsible for a specific task, can be developed, tested, and deployed independently. This significantly reduces development time and complexity, allowing for parallel development efforts and easier maintenance of larger applications. Consider the benefits of reusing a well-tested component in multiple projects; this inherent scalability is one of COM's biggest strengths, allowing for flexibility and efficiency in scaling development efforts. Efficient software development, therefore, hinges on leveraging this powerful framework.

One key aspect of COM is its reliance on interfaces. Instead of directly manipulating a component's internal structure, client applications interact with it solely through these defined interfaces. This abstraction hides implementation details, allowing for changes within a component without affecting the client code, as long as the interface remains consistent. This ensures forward compatibility; components can be updated or replaced without impacting applications using them. The power of this approach lies in its robustness and ease of maintenance, allowing for continuous improvement of individual components without disrupting the larger application. The importance of meticulous interface design cannot be overstated; it's the foundation of a well-structured and maintainable COM-based system.

To effectively utilize COM, a strong grasp of the concept of reference counting is essential. Each component maintains a count of how many clients are currently using it. When this count reaches zero, the component can safely release its resources. This automatic resource management is critical for preventing memory leaks and ensuring the stability of the application. Failing to properly manage reference counts can lead to instability or crashes, highlighting the importance of mastering this fundamental aspect of COM programming. Understanding reference counting is not merely helpful; it's crucial for preventing common pitfalls and building robust, reliable components. Therefore, diligence in managing reference counts should be an integral part of any developer's workflow.

Building COM components often involves using Interface Definition Language (IDL). IDL allows you to define the interfaces your component will expose in a language-neutral way. This definition is then used by compiler tools to generate the necessary code for creating and using your component. Using IDL promotes consistency and facilitates the development of interoperable components across different programming languages. The ability to build components that seamlessly integrate across multiple languages is a significant advantage of this structured approach. It greatly simplifies the integration between components written in different languages, expanding the scope and potential of your development efforts. It removes the language barrier between components, leading to a more unified and efficient system.

COM offers a powerful mechanism for managing component lifetimes and resources, namely, the use of QueryInterface. This method allows a client to query a component for support of specific interfaces. If the component supports the requested interface, it returns a pointer to it; otherwise, it returns an error code. This dynamic interface resolution enhances the flexibility and reusability of components. QueryInterface allows applications to work with components in a flexible manner, adapting to the different interfaces a component might expose, improving adaptability in diverse application contexts. Its dynamic nature is integral to COM's flexibility. Mastering this function is a significant step toward becoming proficient in COM development.

1.2 Interface Definition

Defining interfaces is the cornerstone of successful COM component construction. A well-defined interface acts as a contract, clearly specifying the functionality offered by your component. This clarity is crucial; it prevents misunderstandings between the component and its clients, ensuring seamless integration and reducing the likelihood of errors. Think of it as a meticulously crafted blueprint—a detailed plan that everyone understands and adheres to. Neglecting this crucial step can lead to unforeseen problems, significant debugging challenges, and ultimately, a fragile and unreliable system. A poorly defined interface is akin to building a house without a proper architectural plan; you might get something standing, but its stability and functionality will likely be severely compromised.

Consider the importance of clearly defining the methods, their parameters, and their return values. Every detail matters, every parameter type should be explicit. Ambiguity leaves room for error and misinterpretation, leading to potential crashes, unexpected behavior, and frustrating debugging sessions. Thoroughly document each aspect of your interface. This documentation doesn't merely serve as an internal guide; it becomes a crucial part of your component's public image, aiding other developers in understanding and properly utilizing your creation. Remember that a well-documented interface significantly reduces the learning curve, encouraging broader adoption and greater potential impact.

The interface definition language, or IDL, plays a vital role in this process.

IDL isn't just a programming language; it's a powerful tool that allows you to declare your interfaces in a standardized and platform-independent manner. Think of it as a universal translator, bridging the gap between different programming languages and operating systems. Using IDL allows you to specify your interface once, and then have it automatically translated into different programming language-specific implementations. This flexibility is essential for creating truly reusable and interoperable components, broadening their appeal to a wider developer community and improving the chance of wider adoption.

Choosing appropriate names for your methods and interfaces significantly impacts readability and maintainability. Employ descriptive and self-explanatory names. Avoid cryptic abbreviations or jargon; clarity and consistency are paramount. A well-named interface makes your code more understandable and maintainable, reducing the effort required for debugging, modification, and future development. In essence, thoughtfully chosen names act as clear and concise signposts for developers navigating your code, reducing misunderstandings and enabling easier collaboration.

Error handling is another critical aspect to consider when defining your interface. Decide how your component will manage and report errors. Will it throw exceptions? Will it return specific error codes? Regardless of your chosen approach, consistency is crucial. A consistent error-handling strategy enhances maintainability and predictability, reducing the likelihood of errors related to unpredictable component behaviors. Implementing a clear and efficient error-handling strategy significantly improves the robustness of your component, leading to a more stable and reliable system. Without this step, a minor mishap could snowball into a major system failure.

Beyond the basic method declarations, consider broader design aspects. Will your interface support inheritance? Will you use aggregation or containment to manage complex relationships between components? Carefully planning these features during the interface definition phase prevents later redesigning, significantly reducing development time and effort. It is akin to creating a carefully drafted architectural plan before breaking ground on a building; it ensures the structural integrity of your component, enhancing its adaptability and longevity. Ignoring this aspect can lead to significant structural issues and limit your future development options.

Finally, meticulous testing is indispensable. Once you've defined your interface, thoroughly test it. This comprehensive testing should explore all possible scenarios, handling both expected and unexpected inputs. Verify that your component behaves as expected, that errors are handled gracefully, and that the interface performs efficiently. This step is crucial; it safeguards against unexpected issues, ensuring the overall reliability and stability of your system. Thorough testing saves you from costly debugging sessions and improves overall software quality. Remember, a well-tested interface is a reliable interface, and a reliable interface forms the backbone of a successful COM component.

2 Component Design Principles

Designing effective COM interfaces is paramount. Avoid overly broad interfaces; instead, favor smaller, well-defined interfaces that address specific functionalities. This promotes code reusability and reduces coupling between components. Think of it like building with LEGO bricks – smaller, specialized bricks allow for far greater flexibility and creative construction. Overly large interfaces are akin to using massive, unwieldy blocks that limit your design options and make modification difficult. Careful planning in this phase significantly impacts the maintainability and scalability of your application. Consider the principle of least privilege, granting components access only to the resources absolutely necessary for their operation. This minimizes potential security vulnerabilities and simplifies debugging. Remember that a well-structured interface anticipates future needs. Allow for extensibility through inheritance and polymorphism, ensuring your components can adapt to evolving requirements without requiring substantial redesign. This future-proofing approach is crucial for long-term software sustainability. Invest time in thoroughly documenting your interfaces. Clear and concise documentation is invaluable for other developers – or even your future self – to understand how to interact with your components. Consistent naming conventions are essential for readability and maintainability, establishing a clear pattern for understanding the purpose of each method and property.

Aggregation and containment are powerful techniques for managing relationships between COM components. Aggregation allows one component to use the functionality of another without owning it. This is analogous to hiring a subcontractor – you use their expertise but don't directly control their internal workings. It fosters loose coupling and promotes code reuse. Containment, on the other hand, involves one component fully owning another. This resembles a parent-child relationship, where the parent component has complete control over the child's lifecycle. Choosing between aggregation and containment depends on the specific relationship between the components. Aggregation is generally preferred when a component needs access to the functionality of another without tight coupling, whereas containment is suitable when a component manages the lifecycle of another. Carefully consider the implications of each approach for maintainability and performance. Overuse of either can lead to design complications. Striking a balance between these two approaches is key to building a robust and efficient component architecture. For example, imagine a word processor. The spell-checking functionality might be implemented as an aggregate component, while a font rendering engine might be contained within the main application. These different approaches allow for flexibility in design and maintainability.

Robust error handling is crucial for creating reliable COM components. Never let exceptions silently propagate – always handle them gracefully. This involves implementing comprehensive try-catch blocks to catch and manage potential errors. The key is to provide informative error messages to the caller, allowing them to recover from errors or take appropriate action. Avoid simply displaying generic error messages; instead, provide specific and detailed information about the cause of the error. This helps simplify debugging. This level of detail is especially crucial in distributed systems, where diagnosing the root cause of an error can be challenging. Consider using standardized error codes, making it easier to categorize and handle errors consistently across your application. A consistent approach simplifies debugging, making it easier to track the origins and flow of errors. Moreover, a well-designed error-handling system enhances the overall user experience by providing constructive feedback to users when problems occur, rather than leaving them in the dark. This proactive approach is vital for building dependable and user-friendly software. Remember to always log errors, providing detailed information about the context of the failure and any relevant data. A comprehensive error log becomes an invaluable resource during troubleshooting and allows for effective analysis of the software's behavior. This practice of logging is essential for identifying trends and patterns that may otherwise go unnoticed.

Creating efficient components necessitates careful consideration of resource management. Always release resources promptly, such as COM objects and memory, to prevent leaks. This is especially crucial for long-running applications, where memory leaks can lead to instability and performance degradation. This includes releasing interfaces explicitly, not relying solely on garbage collection mechanisms. Implementing proper resource cleanup is a core component of software development best practices. Efficient memory management and proper interface release significantly reduce the risk of these issues. The impact of neglecting these practices can result in application crashes, system instability, and considerable difficulties during troubleshooting. Proper cleanup procedures prevent these critical failures and increase the reliability of the component. A well-managed component operates predictably and reliably, which is critical for creating successful applications.

Finally, performance optimization is crucial for building responsive and efficient COM components. Profile your code to identify performance bottlenecks. Optimized code reduces unnecessary processing steps, leading to more responsive applications. Profiling allows you to pinpoint specific sections of your code consuming excessive resources. Use tools and techniques to measure resource utilization and identify where improvements can be made. Once these bottlenecks are identified, you can target them for optimization, focusing your efforts on the areas that have the greatest impact on performance. Avoid premature optimization, focusing instead on well-structured and well-tested code initially. However, once you have a stable foundation, you can refine the performance of your code through various techniques. Efficient algorithms and data structures can greatly improve performance. Consider alternatives that provide faster solutions for complex tasks, thereby improving the overall efficiency of the component. Understanding your application's usage patterns is essential for optimizing performance appropriately.

2.1 Interface Design Best Practices

Designing effective COM interfaces is paramount for creating reusable and robust components. Avoid overly complex interfaces; strive for simplicity and clarity. Each interface should have a clearly defined purpose, focusing on a specific set of functionalities. Think of it like building with LEGOs – smaller, well-defined blocks are easier to manage and combine into larger structures. Overly broad interfaces are difficult to maintain, test, and reuse effectively. Remember the principle of least privilege – grant only the necessary access to each interface. This improves security and reduces the risk of unintended side effects. A well-designed interface is like a well-oiled machine – efficient and predictable in its actions.

A key consideration is versioning. As your application evolves, so will your interfaces. Plan for versioning from the outset. Use GUIDs to identify interfaces uniquely across different versions. This ensures backward compatibility and prevents conflicts when different versions of your components coexist within a system. Thoroughly document all interface changes, highlighting any breaking changes. Careful versioning ensures a smooth upgrade path and prevents compatibility issues. Always remember to test thoroughly after any version change. Failing to account for versioning leads to frustrating integration problems and software instability.

Consider using aggregation and containment to manage component relationships. Aggregation lets you embed one component inside another, providing a way to compose more complex functionalities. This enhances modularity and maintainability. Containment is another powerful technique; it involves simply holding references to other components. Both these approaches have different implications for ownership and lifecycle management. Choose the appropriate method based on the specific needs of your component interactions. Understanding the nuances of aggregation and containment is crucial for building flexible and efficient component architectures. Remember, improper use can lead to resource leaks and unexpected behaviour.

Thorough error handling is vital. Avoid using generic error codes. Instead, create custom error codes that clearly indicate the specific problem. Provide detailed error messages that are informative enough for developers to diagnose the root cause. Throwing exceptions is a good practice, but you should also consider creating a mechanism for returning error codes, especially if your components interact with legacy systems. Remember to handle errors gracefully; a crash in one component shouldn't bring down

the whole application. Robust error handling contributes to a more stable and reliable application. Well-defined error reporting mechanisms are a developer's best friend for debugging and troubleshooting.
Finally, thorough testing is essential. Unit tests help ensure each individual interface functions correctly. Integration tests are crucial for verifying that multiple components interact seamlessly. Regression tests are equally important, to make sure updates haven't introduced unintended consequences. Testing should be an integral part of your development process. It's not just a final step – it's an ongoing process that builds confidence in the stability and reliability of your software. Don't skimp on this stage; inadequate testing is the root of many software failures. Invest time in robust testing frameworks and methodologies.

2.2 Aggregation and Containment

Aggregation, in the context of COM, offers a powerful mechanism for building complex components from simpler ones. Imagine you're constructing a sophisticated multimedia player. You might have separate components for audio decoding, video rendering, and subtitle display. Aggregation allows you to seamlessly integrate these individual components into a single, cohesive player without requiring extensive code modifications or significant performance overhead. The key here is that the aggregated components maintain their own identities, but their functionality is exposed through the containing object's interface. This structure promotes modularity and reusability, making maintenance and upgrades significantly easier. For example, swapping out a video decoder for a newer, more efficient one only requires replacing the aggregated component; the main player remains unaffected, demonstrating a level of flexibility that many other integration methods lack. This is a crucial aspect of efficient software design, enhancing not only maintainability but also long-term adaptability in the fast-paced world of software development.

Containment, on the other hand, presents a different approach to component integration. Instead of exposing the aggregated components' functionality directly, the container holds and manages them, providing its own interface for interaction. This is akin to a car's engine; the engine exists independently, but its functionalities are accessed and controlled via the car's steering wheel, pedals, and other interface elements. The contained component's interface is not directly accessible to the outside world. This architectural choice offers a higher degree of encapsulation, protecting the internal workings of the contained component from external access, which could lead to unexpected behavior or security vulnerabilities. This approach improves the robustness of the system. Consider a word processor; the spell-checking functionality might be a contained component. The user interacts with the spell-checker through the word processor's interface, not directly with the spell-checking component itself. This separation improves the integrity and maintainability of the overall system, reducing risks related to unintended consequences and simplifying the development process.

Choosing between aggregation and containment is a critical design decision. Aggregation is ideal when you need to expose the functionality of multiple components directly to the user or another component, ensuring flexibility and code reusability. This is particularly useful when building extensible applications, where adding new features often requires integrating new components into an existing system. Consider the numerous plugins available for popular image-editing software, each offering its own set of specialized tools that the software integrates without compromising the user interface or the overall stability of the application. Such flexibility is pivotal for the success of many modern software systems.

Conversely, containment is preferable when you want to encapsulate the complexities of a component, safeguarding it from external interference and simplifying the overall architecture. This is often favored when dealing with complex or sensitive operations where the potential for errors is high or security concerns are paramount. For example, secure data processing in financial applications benefits significantly from containment, ensuring that the process happens within a controlled environment which prevents unauthorized modification or access to sensitive data.
Effective usage of aggregation and containment significantly impacts the overall design of your COM components. A thoughtful selection, based on the specific requirements of the application and components being integrated, is key to building robust, maintainable, and scalable software systems. The proper implementation of these techniques reduces complexity and promotes efficiency throughout the entire software development lifecycle. Consider the long-term implications; choosing incorrectly can lead to increased maintenance costs and future development difficulties that would otherwise be avoided through a careful initial design choice and diligent implementation of this COM fundamental technique. Failing to understand the nuances of these concepts can result in software that is brittle, difficult to maintain, and ultimately, far more costly to operate in the long run. Therefore, a proper understanding of the differences between aggregation and containment, and their individual strengths and weaknesses, is essential for success in advanced COM component development.

2.3 Error Handling Strategies

Proper error handling is paramount in COM component construction. Neglecting this crucial aspect can lead to application crashes, data corruption, and ultimately, user frustration. Robust error handling ensures the stability and reliability of your components, a critical consideration for any professional application. Think of it as building a sturdy bridge – neglecting a single support beam can cause the whole structure to collapse.

Effective error handling isn't just about catching exceptions; it's about gracefully handling unexpected situations and providing informative feedback. This prevents your component from abruptly terminating, preserving valuable data and guiding users towards a solution. Imagine a user attempting a complex operation. A poorly-handled error could lead to data loss and leave them completely bewildered. A well-handled error, however, will guide them and prevent this catastrophic failure.

The cornerstone of effective error handling in COM is the use of HRESULTs. These 32-bit values encode both success and failure information, providing context to the error. Rather than relying on generic exceptions, HRESULTs enable you to deliver precise error codes to your calling applications. This specific information allows for sophisticated error handling on the client side; they can interpret the code and respond accordingly. A vague error message only tells half the story; a specific HRESULT gives the whole picture.

Consider using standard HRESULTs wherever possible. This promotes interoperability and allows other developers to readily understand the errors your component might raise. Consistency fosters easier debugging and reduces integration issues. Imagine the frustration if each component uses a completely different error code system. This adherence to standards will simplify your interactions immensely.

Beyond standard HRESULTs, you should also meticulously document all custom error codes your component may generate. Clear documentation is invaluable for maintaining your component, supporting its users, and allowing for effective collaboration. Think of your code as a conversation:

Clear communication is essential for mutual understanding. Each custom error needs a helpful explanation to make sense of the situation.

Never underestimate the value of informative error messages. While HRESULTs provide the technical details, accompanying error messages should provide clear, user-friendly explanations of the problem. Avoid technical jargon and instead aim for concise, easily understood language. A cryptic error message only adds to the confusion, while a clear one can lead to quick problem resolution. Aim for the clear language that all the developers would understand, no matter their knowledge.

Implementing comprehensive error logging within your component is another critical aspect. This provides a detailed record of errors that occur during runtime, invaluable for debugging and post-mortem analysis. Consider logging the timestamp, HRESULT, error message, and any relevant context information such as inputs and internal state. Log files will provide a crucial help in debugging.

Implementing error recovery mechanisms is a significant factor in robust error handling. Where possible, your component should attempt to recover from errors gracefully, preventing cascading failures. A resilient component can potentially correct issues on its own, minimizing disruption to the application. This proactive approach shows foresight and minimizes negative impacts.

Testing your error handling is as vital as implementing it. Employ rigorous unit and integration testing to simulate various failure scenarios. Ensure your error handling mechanisms function as expected, reporting errors accurately and recovering gracefully. Consider edge cases, boundary conditions, and other situations where problems are more likely to occur.

Proper error handling isn't a mere afterthought; it's an integral aspect of secure, robust COM component design. By carefully employing the strategies discussed, you can significantly improve the reliability and maintainability of your components, ensuring a positive experience for both developers and end-users alike. It's about building software that is not just functional, but also resilient and user-friendly. This commitment to quality elevates your work from good to great.

3 Advanced COM Features

Marshaling is the process of transferring data between different memory spaces, often across process boundaries. Understanding this is crucial for building robust COM components. For instance, consider a scenario where a client application needs to pass a complex data structure to a component residing in a separate process. Simple data types, such as integers or strings, are relatively straightforward to marshal. However, more intricate data structures require a more nuanced approach. The COM library offers built-in marshaling for standard data types, relieving you from the lower-level details. However, when dealing with custom data structures, you need to define how they're marshaled, using custom marshaling techniques we'll explore later in this chapter. Proper marshaling ensures data integrity and prevents unexpected errors. Failing to do so can lead to crashes or data corruption, severely impacting your application's stability. Ignoring this fundamental aspect can result in hours of debugging and frustration. Therefore, meticulous planning regarding marshaling is critical to success. Security is paramount, particularly when dealing with components that might access sensitive data or resources. COM provides several mechanisms for enhancing component security, but you must actively employ them. One vital aspect involves access control. You must carefully define which users or processes have permissions to access and utilize your component's functionality. This is achieved using various security mechanisms, such as role-based access control or more granular permission settings. Failure to secure your component adequately exposes it to potential security breaches. Malicious actors could exploit vulnerabilities, compromising your application's integrity and potentially exposing sensitive information. Moreover, poorly secured components can become entry points for malware, causing significant damage to the system. Thorough security planning and implementation are vital for building secure, reliable applications. Never underestimate the importance of security; it's not merely an afterthought but an integral design element.

Event handling in COM involves asynchronous communication between components. Imagine a scenario with a component generating events, like a sensor sending readings. Other components need to react to these events without blocking the event producer. COM offers mechanisms like connection points and event sinks, facilitating this asynchronous interaction. Connection points serve as interfaces that the event producer exposes to interested clients. Event sinks act as handlers in the client components that process these events. The interaction adheres to a publisher-subscriber model where the producer publishes events, and clients subscribe to receive them. Efficiently handling these events is essential for responsiveness and scalability. Inefficient handling can cause delays or missed events, leading to application malfunction or data loss. Proper event handling is critical for building responsive and robust COM applications. This is a fundamental concept that separates well-designed components from unstable ones.

Threading models determine how your component interacts with different threads. In multi-threaded environments, careful consideration of the threading model is vital for avoiding race conditions and deadlocks. COM offers different threading models, including apartment-threaded, free-threaded, and both. Understanding their implications on your component's design is crucial. For example, the apartment-threaded model, where each component resides in a single thread, simplifies concurrency management but limits performance in heavily threaded applications. Free-threaded models offer better performance and scalability but demand meticulous synchronization mechanisms to ensure data integrity. Choosing the correct threading model significantly impacts the performance, stability, and scalability of your component. A poorly chosen threading model can directly lead to crashes and application instability. You must understand the implications of each model before making a decision.

Custom marshaling is used when the standard marshaling techniques provided by COM are insufficient. This often occurs when dealing with complex data structures or custom interfaces. Building custom marshalers requires a deep understanding of COM's internal mechanisms. You'll need to write code that explicitly defines how your custom data structures are transferred between process boundaries. Effective custom marshaling ensures that your custom data is correctly marshaled and unmarshaled, preventing corruption or errors. However, improper implementation can lead to crashes, data loss, and significant debugging challenges. Mastering this aspect allows you to create components capable of handling highly specialized data, enhancing flexibility and interoperability. This is a powerful tool, but its correct implementation is crucial for application stability. Remember to thoroughly test your custom marshaler to avoid unforeseen problems.

3.1 Marshaling Techniques

Marshaling, at its core, is the process of transferring data between different memory spaces. This is particularly crucial in COM, where components might reside in separate processes, even on different machines. Understanding the intricacies of marshaling is fundamental to building robust and efficient distributed applications. Failing to grasp these mechanics can lead to unexpected crashes, data corruption, and significant performance bottlenecks. Therefore, a deep understanding is not merely beneficial, it's essential.

Consider a scenario where your COM component, running in one process, needs to pass a complex data structure to another component in a different process. This seemingly simple task transforms into a complex dance of data serialization, transfer, and deserialization. The marshaler acts as the choreographer, orchestrating this intricate process to ensure the data arrives safely and intact at its destination. This seemingly straightforward process hides a multitude of complexities, often resulting in unexpected issues if not properly managed. The specifics depend heavily on the type of data being marshaled and the communication mechanism used.

The simplest form of marshaling involves basic data types—integers, floating-point numbers, and strings. These are typically handled efficiently by the COM runtime itself. However, more complex data structures, such as custom classes or arrays of objects, demand more sophisticated strategies. Here, you need to define how these custom types should be represented in a serialized format before transmission, a format that can then be faithfully reconstructed on the receiving end. This requires meticulous attention to detail, even small oversights can lead to subtle but devastating errors. This intricate process requires a very deep understanding of serialization and deserialization.

For custom data structures, you might need to employ custom marshaling. This involves writing specific code to define how your custom types are represented during marshaling. This gives you ultimate control over the process but necessitates a significant investment in understanding the intricacies of the COM marshaling API. It's a powerful tool but demands meticulous planning and coding, with inadequate attention leading to more errors. You are essentially defining a bespoke data translation process.

One common approach is to use the VARIANT type, a versatile container capable of holding various data types. While convenient, relying heavily on VARIANT can lead to performance degradation, especially when dealing with large amounts of data. Remember to always opt for the most efficient data representation for your specific needs. Over-reliance on this type sacrifices potential optimization. If your architecture requires a large data transfer, consider implementing a more specialized solution for better performance and maintainability.

Choosing the right marshaling mechanism significantly impacts performance. Using inefficient methods can dramatically slow down inter-process communication. Benchmarking your code and comparing different techniques is crucial. In some cases, optimization strategies may improve performance by orders of magnitude. Profiling and performance analysis should be an integral part of the development process. If your component interacts with other processes frequently, this step is critical.

Remember, improper marshaling can lead to security vulnerabilities. If not carefully handled, attackers could exploit weaknesses in your marshaling code to inject malicious data into your system. Secure coding practices, rigorous testing, and a deep understanding of potential vulnerabilities are essential. Implementing robust security checks is paramount. It's not just about efficiency; it's about safeguarding your application.

Finally, debugging marshaling issues can be challenging. The problem may not always manifest in the component where the error originates. Thorough testing and logging are vital to identify and resolve these hidden problems. A robust approach to debugging includes the careful use of logging and tracing to pinpoint where the data transfer fails. This is often a complex debugging effort. Effective use of debugging tools is also essential.

In conclusion, mastering marshaling techniques is a critical skill for any COM developer. While the initial learning curve might be steep, the rewards in terms of building robust, efficient, and secure distributed applications are substantial. It's a skill that separates good COM developers from great ones. Remember this skill is essential to a successful component and application. The time invested in mastering these concepts yields significant returns in the long run.

3.2 Security Considerations

Security in COM component construction is paramount; neglecting it can expose your applications to vulnerabilities. A robust security model protects against unauthorized access, code injection, and data tampering. This involves careful consideration of several key areas, starting from the design phase itself.

Building secure COM components requires a layered approach. First, you must define clear access controls, specifying which clients can access specific methods and properties of your components. This might involve using security descriptors to control access rights, limiting functionality based on the user's identity, or employing role-based access control mechanisms to grant permissions granularly. Proper implementation ensures only authorized entities can interact with sensitive data or functionalities.

Next, input validation is crucial. Never trust the data received from external sources. Always rigorously validate all inputs, checking for data types, length constraints, and potential malicious content. This step should incorporate extensive checks to prevent buffer overflows, format string vulnerabilities, and other injection attacks. Think of it as a protective shield, ensuring only properly formatted and safe data reaches your component's internal workings.

Furthermore, consider the use of strong cryptography. If your components handle sensitive information, like passwords or financial data, encrypting this data both at rest and in transit is essential. Utilize industry-standard cryptographic algorithms and key management practices to protect data confidentiality and integrity. This should not be an afterthought; it's an integral aspect of secure design, crucial to protecting your data.

Moreover, code signing adds a crucial layer of security. Signing your components digitally with a trusted certificate assures clients that the code originates from a verified source and hasn't been tampered with. This helps prevent malicious code from being substituted for legitimate components, reducing the risk of attacks that rely on code manipulation. Thorough validation of certificates ensures authenticity and prevents fraudulent components from being installed.

Furthermore, robust error handling is not just about graceful degradation; it's about security. Improper error handling can leave your components vulnerable to exceptions that might expose sensitive information or allow attackers to exploit vulnerabilities. Handle errors carefully, avoiding the leakage of sensitive data in error messages, and ensure that exceptions are caught and managed safely without compromising the application's security posture.

Beyond individual components, secure deployment practices also contribute greatly. Avoid installing components with excessive privileges, only granting the minimum necessary access rights. Keep your components and their dependencies updated regularly to patch security vulnerabilities and prevent exploitation. Regularly reviewing and updating security policies and practices is critical to maintain a secure environment.

Finally, thorough testing is the cornerstone of secure development. Conduct rigorous testing, including penetration testing and security audits, to identify and address vulnerabilities before deployment. Simulate real-world attacks to identify weaknesses and improve your security posture proactively. This iterative process will strengthen your components and bolster their resilience against potential threats. Remember, comprehensive testing isn't a luxury; it's a necessity for robust and secure COM components. Failing to do so can result in significant security breaches and substantial costs. Proactive security measures save you time, money, and potential reputational damage. It's better to be prepared than to react to vulnerabilities.

3.3 Event Handling Mechanisms

Efficient event handling is paramount in robust COM component design. Ignoring this critical aspect can lead to unresponsive applications and unpredictable behavior, undermining the very advantages of using COM for modularity and reusability. Properly implemented event handling allows components to communicate asynchronously, enhancing responsiveness and overall application performance. A well-structured event handling system ensures smooth data flow between components, even under heavy load.

Consider a scenario where a data acquisition component needs to notify a user interface component of new data arrivals. A simple polling mechanism would be inefficient, consuming significant resources and introducing latency. Instead, an event-driven approach, where the data acquisition component raises an event upon data arrival, allows the user interface to react immediately and only when necessary. This asynchronous communication is fundamental to building responsive applications.

The cornerstone of COM event handling lies in the concept of source and sink interfaces. The event source, typically the component generating the event, exposes an interface that allows clients to subscribe to specific events. The event sink, on the other hand, is an interface implemented by the client to receive notifications. This clear separation of concerns promotes modularity and maintainability. Through this well-defined structure, you can implement complex interactions between components without tight coupling, maintaining the flexibility and independence of each component.

Connection points form the mechanism through which the event source and sink interact. The event source maintains a list of subscribers and uses connection points to marshal events to each registered sink. This architecture facilitates efficient event distribution, regardless of the number of subscribers. Effective use of connection points requires a thorough understanding of IConnectionPoint and IConnectionPointContainer interfaces. Careful design of these interfaces is crucial for establishing a stable and scalable event system within your COM components. Proper handling ensures events are consistently distributed, avoiding common pitfalls such as memory leaks or race conditions.

Implementing custom events necessitates careful attention to detail. Defining custom events often involves creating a new interface that inherits from IDispatch or a similar interface, thus providing a mechanism for clients to handle events. It allows components to communicate highly-specific data changes. Methods for raising and handling events must be meticulously crafted to ensure type safety and reliability. Effective use of custom events greatly enhances the flexibility of your components. It opens the door to finely tuned communication, increasing efficiency and responsiveness.

While connection points provide a robust mechanism for event handling, understanding their limitations is equally crucial. The sheer number of subscribers can impose a performance overhead. In scenarios with a large number of subscribers, optimizing the event notification process is paramount. Techniques like event aggregation, or perhaps using a message queue instead of direct notification, can help mitigate potential performance bottlenecks. These strategies are important when balancing efficient event handling and component performance, specifically in resource-constrained environments. Understanding and planning for these considerations is as essential as the initial implementation of connection points.

Debugging event handling mechanisms presents unique challenges. Tracing the flow of events across multiple components can be tedious. Utilizing tools such as debuggers and logging mechanisms is invaluable in isolating and resolving issues within the event handling system. Effective logging practices are critical for understanding event sequences. By logging key events and associated data, you can gain valuable insights into the system's behavior, which is critical for effective debugging. Thorough testing under various scenarios, including multiple subscribers and high event rates, ensures the robustness and stability of your event handling implementation. This is essential for building reliable components.

Finally, remember that security is inextricably linked to event handling. Uncontrolled event subscriptions can create security vulnerabilities, particularly in distributed systems. Implementing access controls and carefully managing event subscriptions are paramount for safeguarding your components. This is critical in scenarios involving sensitive data. Robust security measures, implemented during both the design and implementation phase, are not an afterthought but an essential part of building secure COM components. They safeguard against unauthorized access and malicious activity. The security design must match the robustness of the architecture.

3.4 Threading Models

Apartment threading, also known as single-threaded apartment (STA) threading, is a model where each thread operates within its own isolated apartment. This ensures that COM objects are accessed sequentially, avoiding race conditions and simplifying synchronization. The primary benefit of this approach is its simplicity; it's inherently thread-safe for objects not explicitly designed for multithreading. However, performance can suffer under heavy loads because only one thread can interact with a particular COM object at any given time. This can become a significant bottleneck in applications that require high concurrency. Consider carefully whether the simplicity outweighs the potential performance implications before choosing this model for your components. You'll want to profile thoroughly to assess the trade-offs if concurrency is critical to your application's functionality. Remember, it's easy to underestimate the impact of thread contention.

A free-threaded marshaler (FTM) offers significantly more flexibility. This model uses an intermediary, the marshaler, to manage communication between threads. It enables multiple threads to concurrently access the same COM object without encountering the limitations of the STA model. The marshaler handles the complexities of thread synchronization, making it easier to develop high-performance, multithreaded components. Although seemingly complex, utilizing FTMs often leads to far more efficient and responsive applications, especially those interacting with many threads. However, proper thread synchronization is crucial when working with FTMs, and you must carefully manage the interactions between threads to prevent data corruption or unexpected behavior. The overhead introduced by the marshaling process should also be considered; while often minor, it's not negligible for very high frequency interactions. Thorough testing under various load conditions is key to verifying the effectiveness of your FTM implementation.

Neutral threading is a powerful and efficient option, allowing both STA and MTA objects to coexist within the same process. This model offers a good balance between simplicity and performance; allowing you to employ both STA components and MTA components according to your needs, making use of the respective strengths while minimizing the weaknesses. This strategy makes your component more adaptable, but also increases the complexity of your design. You'll need to carefully manage the interactions between STA and MTA components, being mindful of potential synchronization issues. The flexibility inherent in the neutral threading model is a significant advantage, but its complexity demands a careful, phased approach to design and thorough testing to ensure robustness. Understanding how different threads interact with different component types is key to successful neutral threading.

Choosing the correct threading model is a critical design decision influencing performance, scalability, and maintainability. For simple components where concurrency is not a primary concern, the apartment-threaded model offers a straightforward approach with minimal complexity. Conversely, for high-performance applications requiring concurrent access to components, the free-threaded marshaler model is usually the preferred choice, although its implementation requires a more robust understanding of thread management and synchronization. Neutral threading provides a powerful compromise. It is not always an ideal choice but it can be the right option for many complex applications. Each model has trade-offs, and the best choice will be dependent on the specific requirements of your application. Always select the option that delivers the best compromise between complexity and performance. Don't simply choose the easiest method unless there are no other demands.

Beyond the fundamental choices, remember to consider factors such as the complexity of your component interactions. A component with a relatively simple interface and limited interaction might be suitable for the simplest threading model. On the other hand, a component involved in complex interactions with many other threads, perhaps involving substantial data transfer, might necessitate a more robust model such as the free-threaded marshaler to prevent performance bottlenecks and ensure thread safety. A thorough understanding of the threading model's implications regarding performance and scalability is essential to make an informed decision during the design phase. Careful design will avoid unforeseen issues down the road. Remember, choosing the right threading model is a foundational step in building efficient and robust COM components; a mistake here can have far-reaching consequences.

Finally, thorough testing is essential to confirm your choice of threading model aligns with the performance requirements of your application. Monitor performance under realistic load conditions to identify potential bottlenecks. Remember that selecting the wrong threading model can lead to significant performance issues, even if your individual components are well-designed. Pay close attention to areas where you find bottlenecks when testing. If you notice performance problems, you might need to revisit your threading model choice, perhaps modifying the design of your components to match the chosen threading strategy better. Consider using profiling tools during testing to understand the performance profile of your components and identify areas for potential optimization. Don't hesitate to iterate on your design as you refine your understanding of threading model performance within your particular application. The result will be far more robust and performant software that meets your needs.

3.5 Custom Marshaling

Custom marshaling offers unparalleled control over how your COM components exchange data. Unlike the default marshaling mechanisms, which handle basic data types effectively but might struggle with complex structures or custom data representations, custom marshaling allows you to define precisely how your data is converted and transferred between processes. This fine-grained control is vital for optimization, especially when dealing with large datasets or unique data formats. You gain the ability to tailor the marshaling process to your specific needs, potentially leading to significant performance improvements.

Implementing custom marshaling involves creating custom marshaler classes. These classes inherit from the IMarshal interface and override its methods to manage the data transfer. This is more involved than relying on default marshaling, demanding a deeper understanding of COM's inner workings. However, the payoff is substantial, especially when dealing with intricate data structures or legacy systems that require unique handling. You'll need to carefully design your marshaler to handle different scenarios like data serialization and deserialization across process boundaries, ensuring robustness and error handling.

Consider a scenario where your component interacts with a legacy system using a specialized data format. Default marshaling would likely fail to handle this format correctly, leading to data corruption or application crashes. With custom marshaling, you can write a marshaler specifically designed to convert your component's internal data representation to the format understood by the legacy system and vice versa, ensuring seamless interoperability. This becomes even more important when working with complex data structures, such as custom classes containing arrays, pointers, or other intricate elements. A well-designed custom marshaler guarantees data integrity.

The process of creating a custom marshaler requires a systematic approach. You begin by defining the data structures that need custom marshaling. This necessitates a thorough understanding of both your component's internal data structures and the external systems it needs to interact with. Next, you implement the IMarshal interface methods, meticulously crafting the logic for converting data between these representations. Remember to thoroughly test your custom marshaler under various conditions, including edge cases and error scenarios, to guarantee reliability. This meticulous testing ensures that your custom marshaler functions flawlessly in different environments and under stress.

One significant advantage of custom marshaling lies in its optimization potential. By carefully designing your marshaler, you can minimize the amount of data transferred between processes. This is especially crucial in scenarios where bandwidth is limited or where the data involved is exceptionally large. You might employ techniques such as data compression or selective data transfer to reduce the overall data footprint. This level of optimization is simply not possible with the standard, more generic marshaling methods. Efficient custom marshaling can drastically improve the responsiveness and performance of your applications.

Beyond performance enhancements, custom marshaling also strengthens security. By controlling the marshaling process, you can incorporate security checks and validations into the data transfer process. This can involve validating the data's integrity, ensuring that it hasn't been tampered with during transmission. Furthermore, you can implement access controls to restrict which data is marshaled, enhancing the overall security of your component and the applications that use it. This proactive approach to security is crucial in today's data-centric world.

However, custom marshaling demands meticulous attention to detail and thorough testing. Errors in the custom marshaler can lead to unexpected behavior, data corruption, or even system crashes. It is essential to meticulously debug your code and comprehensively test your custom marshaler in various scenarios to guarantee its correctness. This rigorous testing approach minimizes the risk of unforeseen problems, ensuring the reliability and robustness of your COM component. Remember that the complexity of custom marshaling might outweigh the benefits in simpler cases; only consider it when the standard marshaling methods prove inadequate.

Debugging custom marshalers can be challenging. Traditional debugging techniques might not suffice, demanding a deeper dive into the COM marshaling infrastructure. Using debugging tools that allow you to inspect the data during the marshaling process can be invaluable, allowing you to trace the flow of data and identify potential problems. Thorough logging and error handling within the marshaler itself will aid in detecting and resolving issues. The effort invested in robust debugging will pay off in the long run, ensuring a stable and reliable component.

In conclusion, while implementing custom marshaling is more complex than using the default mechanism, it provides significant advantages in terms of control, optimization, and security. By carefully crafting your custom marshaler and meticulously testing it, you can achieve superior performance and reliability, especially when dealing with complex data structures or interfacing with legacy systems. The ability to fine-tune the data transfer process gives you an edge in building high-performance, secure, and robust COM components. However, always weigh the added complexity against the benefits before undertaking this more advanced approach.

4 Building Efficient Components

Efficient COM component construction hinges on meticulous attention to performance and memory management. Ignoring these aspects can lead to sluggish applications, resource leaks, and ultimately, system instability. Therefore, proactive measures are paramount.

Optimizing performance requires a multi-pronged approach. Profile your code to identify bottlenecks. Tools like VTune Amplifier or similar profilers pinpoint performance-critical sections. This allows you to focus your optimization efforts on areas with the greatest impact. Don't waste time on already efficient code. Once you've identified bottlenecks, focus on algorithm optimization. Can you use more efficient algorithms or data structures? For example, replace slow linear searches with quicker binary searches where applicable. Consider using appropriate data structures to reduce processing time, and remember to measure the impact of any changes. This iterative approach ensures your efforts are effective.

Furthermore, minimizing the number of COM calls is crucial. Each COM call involves a degree of overhead, so minimizing them improves overall responsiveness. This often involves careful design of the component's interfaces. Aim for broader, more comprehensive interfaces that consolidate multiple individual calls into fewer, combined actions. This reduces the cross-process communication overhead. Aggressively caching frequently accessed data can also provide significant performance improvements. By storing results in memory, you avoid redundant computations. In the same vein, use asynchronous operations strategically. While they add complexity, asynchronous calls allow for parallel processing, thus avoiding unnecessary waiting times. This is especially beneficial in I/O-bound operations.

Memory management in COM is crucial. Improper memory handling directly leads to memory leaks, eventually causing system instability. The primary culprits are often forgotten releases of interfaces. Always meticulously release interfaces using the Release() method when you're finished with them. Avoid unnecessary object creation. Create objects only when needed, and destroy them as soon as they're no longer required. This prevents the accumulation of unneeded memory. Consider using smart pointers. They provide automatic resource management. This eliminates the need for explicit Release() calls, thus reducing the risk of errors. While seemingly simple, these practices, rigorously applied, make a substantial difference.

Consider implementing reference counting effectively within your components. This provides a robust mechanism for managing the lifetime of your objects. Ensure your reference counting is implemented flawlessly. Errors in reference counting can have severe consequences. Testing and thorough review of your component's memory management are essential before deployment. Memory leaks, particularly in long-running applications, become progressively worse, consuming system resources and potentially causing crashes. Therefore, rigorous testing is not merely advisable, it's essential. Testing with different data volumes and usage patterns reveals hidden problems before they impact end-users. Memory debuggers, alongside carefully crafted unit and integration tests, are invaluable assets in identifying and

resolving such problems. Investing time in thorough testing is an investment in stability and reliability.
Beyond the individual components, consider the overall system architecture's impact on efficiency. Careful design of component interactions can significantly enhance performance. Minimizing unnecessary inter-component communication reduces overhead. Therefore, efficient system architecture significantly contributes to the overall performance of your application. Using suitable tools to detect performance issues and bottlenecks early in the development process is crucial. These strategies allow for early detection and resolution of performance issues, saving time and effort in the long run. By integrating these principles into your development workflow, your components will be more efficient, reducing resource consumption and improving the overall user experience. Remember, efficiency is not a single technique, but a comprehensive approach.

4.1 Performance Optimization

Efficient COM component performance hinges on several key areas. Careful interface design, discussed previously, directly impacts speed; minimizing the number of method calls and the data transferred through interfaces is crucial. Consider using aggregation where appropriate to avoid unnecessary object creation and destruction, leading to faster execution times. Remember, poorly designed interfaces can lead to performance bottlenecks you'll spend hours chasing down later.
Memory management is paramount. COM components must meticulously manage their resources, avoiding memory leaks which, if left unchecked, can degrade performance and lead to application crashes. Always explicitly release interfaces using Release() when they are no longer needed. Failure to do so can significantly impact performance, especially in long-running ap-

plications. Implement robust error handling, as unexpected errors can lead to resource contention and performance degradation. This means properly handling exceptions and cleaning up after errors to ensure resources are freed, preventing potential memory issues.

Understanding the implications of marshaling is vital for efficient component interaction. Marshaling, the process of transferring data between processes, can introduce significant overhead if not managed carefully. Choosing the appropriate marshaling type, based on the size and type of data, is essential for optimizing performance. Favor in-process components whenever feasible, as they avoid the overhead associated with cross-process communication. For out-of-process components, explore options like using lightweight marshaling techniques to reduce data transfer costs.

Threading models directly affect performance, especially in multithreaded environments. Choosing the right threading model for your component, whether single-threaded apartment (STA), multi-threaded apartment (MTA), or both, is critical. Properly managing thread synchronization can prevent deadlocks and race conditions that can drastically slow down your component. Consider using asynchronous operations to improve responsiveness and avoid blocking the main thread, thereby maintaining high performance even under heavy load.

Profiling your components is a crucial step in performance optimization. Use profiling tools to identify bottlenecks in your code. This allows you to pinpoint areas where your component spends the most time. Once you've identified these bottlenecks, you can focus your optimization efforts on the most impactful sections of your code. This targeted approach ensures that your optimization efforts yield the maximum performance improvements.

Code optimization techniques should be applied strategically. Avoid premature optimization; focus first on improving the overall design of your component before delving into micro-optimizations. Using efficient algorithms and data structures can significantly improve performance. Consider techniques like memoization to cache frequently accessed results, and minimizing unnecessary calculations. Profiling will guide you to the most effective areas for optimization.

Remember the importance of using the correct data types. Employ data types appropriately to reduce memory usage and improve processing speed. Larger data types might improve precision, but at the cost of performance. Choosing the most efficient data type for each variable can significantly impact the overall performance of your component. Use profiling to determine if any type changes significantly impact the performance.

Careful consideration of debugging techniques and thorough testing are essential to achieving optimal performance. Debugging tools allow for identifying performance issues. Thorough testing is needed to validate that the applied optimizations truly improve performance and have not introduced new bugs. Regress testing after each optimization iteration can help detect unexpected behaviors or performance regressions. This iterative process ensures the performance gains are stable and reliable. Remember that optimization is an iterative process—profile, optimize, test, repeat—until performance goals are met. It is a cycle of continuous improvement, not a one-time fix.

Finally, meticulous testing under various conditions is critical. Test your component under realistic load conditions to identify performance bottlenecks that might not be apparent during simpler tests. Stress testing helps to uncover performance issues that might only arise under heavy loads. Employ benchmarking to measure the performance improvements achieved through your optimization efforts, providing quantifiable evidence of your success. Thorough and rigorous testing is as critical to the process of optimization as the implementation of optimization techniques themselves.

4.2 Memory Management

Efficient memory management is paramount in COM component development. Failing to properly handle memory allocation and deallocation can lead to memory leaks, crashes, and unpredictable application behavior. This undermines the reliability and performance of your components, impacting the overall user experience. Therefore, disciplined coding practices are absolutely essential.

Understanding the COM memory model is your first step. COM employs reference counting, meaning that each component maintains a count of how many clients are using it. When this count reaches zero, the component's resources are released. This differs from garbage collection in some other environments. You must actively manage these references. Improper reference counting is a primary cause of memory leaks in COM.

AddRef and Release are the fundamental methods for managing references. AddRef increments the reference count, indicating another client is using the component. Release decrements the count. When the count drops to zero, the component's destructor is called, freeing its resources. Consistent and correct use of these methods is critical. Overlooking even a single call can lead to a persistent leak. It's helpful to consider each method invocation as a contract between the component and the client; it is a responsibility that requires meticulous care.

Consider the scenario of a component creating other objects. These subordinate objects must also have their references carefully managed. Failure to release these resources leads to memory leaks within the component itself. This cascade effect can be devastating, especially in long-running applications. You will need to adopt best practice techniques to ensure that when one component releases another, the subordinate's reference count is correctly tracked.

Smart pointers are a significant aid in managing this complexity. Smart pointers automatically handle AddRef and Release calls, reducing the chance of errors. Many COM development environments provide smart pointer classes which simplify the process. The use of these simplifies the code, rendering it more readable, easier to maintain, and reduces the likelihood of memory-related issues. They're an invaluable tool, streamlining the memory management process dramatically and reducing potential errors.

Beyond smart pointers, allocating and deallocating memory directly through functions like CoTaskMemAlloc and CoTaskMemFree is a critical aspect of managing memory. Always remember to pair these allocations with proper deallocations and ensure the memory allocated is always released when it is no longer needed. Never assume that the system will automatically clean up after you; this is a recipe for disaster. Each allocation must have a corresponding deallocation, and a careful audit of memory handling is essential.

The concept of ownership is intrinsically tied to memory management. Each object should have a clear owner responsible for its lifetime. If multiple objects claim ownership, you will encounter unpredictable results – perhaps conflicts, and almost certainly memory leaks. This requires careful design decisions. Understanding this crucial principle of object ownership will avoid many potential problems.

Thorough testing is essential. Memory leaks often manifest subtly, appearing only under specific conditions or after extended periods of operation. Therefore, implement rigorous testing strategies including stress tests, load tests, and prolonged usage scenarios. Employing memory debugging tools is also highly recommended. These specialized tools can help pinpoint memory leaks and other memory-related issues more effectively and efficiently. The tools will help you visualize and analyze memory usage patterns that may indicate underlying problems.

Debugging memory issues is frequently tedious, however, the effort is well worth it. Memory leaks can severely degrade application performance or even cause crashes. Systematic debugging is often necessary, requiring a thorough examination of your code and the use of memory profiling tools. These tools monitor memory allocation and deallocation, flagging potential problems that might otherwise go undetected. Pay very close attention to detail in the debugging process.

In summary, efficient memory management in COM component development is not optional; it is crucial for building robust, reliable, and high-performance components. The disciplined use of reference counting, smart pointers, and proper memory allocation/deallocation functions are essential. Don't underestimate the importance of thorough testing and debugging tools. Proactive memory management is not merely a best practice; it's a necessity.

5 COM+ Enhancements

COM+ significantly enhances the capabilities of standard COM, offering features crucial for building robust and scalable enterprise applications. Understanding these enhancements is paramount for leveraging the full potential of COM-based development. This chapter delves into the key features provided by COM+, highlighting their practical applications and benefits.

Transaction management, a cornerstone of COM+, ensures data integrity across multiple components. Imagine a complex banking application; if a transfer fails midway, you'd want the entire operation rolled back. COM+ transactions, utilizing resource managers like databases, guarantee atomicity—all parts of a transaction succeed or none do—preventing data corruption and maintaining consistency. This reliability is achieved through sophisticated coordination of participating components, automatically managing the commit or rollback process. The implementation of transactions typically involves the use of COM+ objects such as the TransactionalContext object to manage transaction boundaries. Proper configuration within the COM+ application and service is also vital for correct and dependable transaction processing. Using these tools allows developers to focus on the business logic rather than low-level transactional details.

Message queuing, another powerful COM+ feature, allows for asynchronous communication between components. This is especially useful in scenarios where immediate responses aren't required or where components operate on different schedules. For instance, consider an e-commerce application: order processing can be queued, allowing the system to acknowledge the order immediately, even if the inventory check and fulfillment happen later. This improves responsiveness and scalability. COM+ integrates smoothly with Microsoft Message Queue (MSMQ) allowing components to efficiently send and receive messages, ensuring reliable and asynchronous data transfer in a loosely coupled fashion. Efficient management of queues and the correct design of message structures are crucial to avoiding performance bottlenecks and ensuring the application's seamless operation. Careful consideration is needed to properly manage queue sizes, error handling within messages, and proper selection of message types to facilitate integration with other system components.

Component services, offered by COM+, provides a sophisticated environment for managing and monitoring COM+ applications. This centralized management simplifies deployment and simplifies the process of configuring security settings and configuring various component parameters. Administrators can easily track application performance, manage component instances, and implement robust security measures from a single, accessible location—enhancing control and overall ease of maintenance. Understanding the options for configuring specific components and monitoring their performance is paramount to ensuring reliable and efficient application functioning.

Effective deployment of COM+ components necessitates careful planning and execution. The COM+ environment includes robust tools and configurations settings for managing, implementing, and deploying COM+ components. Understanding this process is crucial for successful integration

into existing systems. This might involve careful consideration of component versions, dependencies, and the overall application architecture. Furthermore, proper use of deployment tools minimizes potential deployment errors, and facilitates updating components without downtime. Ensuring compatibility with existing systems is equally important for integrating COM+ components without causing any conflicts.
In summary, COM+ offers a significant upgrade to the standard COM architecture. By mastering transaction management, message queuing, component services, and robust deployment strategies, developers can construct highly reliable, scalable, and maintainable enterprise-grade applications. These enhancements simplify complex development tasks and deliver robust solutions capable of handling the demands of modern software systems. The diligent application of the techniques and best practices discussed provides a solid framework for creating sophisticated and efficient software solutions using the COM+ environment. Implementing and leveraging these features correctly enhances the functionality and stability of your applications, producing a superior user experience while simplifying the management process and enhancing overall efficiency.

5.1 Transactions and Queues

Transactions, within the context of COM+, offer a robust mechanism to ensure data integrity across multiple operations. Imagine a scenario involving a banking application; transferring funds requires updating multiple accounts simultaneously. If one update fails, the entire transaction should roll back, preventing inconsistencies. COM+ transactions guarantee this atomicity, ensuring data remains consistent even in the face of partial failures. This is achieved through the use of transaction coordinators which monitor the entire process, ensuring either complete success or complete rollback.

Proper configuration and understanding of transaction boundaries are crucial for leveraging this feature effectively, and improper usage can easily lead to unpredictable behavior and data corruption. Careful planning and design are paramount to avoid such issues. Implementing transactions correctly can significantly enhance the reliability and data integrity of your applications.

Queues, in contrast to transactions, provide a mechanism for asynchronous communication. Instead of directly interacting with a component, a client can place a request in a queue. A separate component, a queue listener, processes these requests independently. This decoupling allows for greater scalability and resilience. A significant advantage of using queues is the ability to handle situations where a component might be temporarily unavailable or overloaded. Requests are stored in the queue until the listener can process them, ensuring no data is lost. This approach is particularly valuable in distributed systems, where communication latency and component failures are common occurrences. Microsoft Message Queue (MSMQ) is a powerful technology for implementing such queue-based systems in COM+ environments. Understanding the nuances of message queuing, including message prioritization and error handling, is key to building robust and responsive applications that effectively manage potentially significant workloads and unexpected failures.

Effective error handling within transactional and queued components requires a different approach than in traditional components. While standard COM error handling mechanisms remain important, anticipating potential failures within transactions and queues is crucial. Consider the implications of a failed transaction: the system must either cleanly roll back or compensate for the partial changes in a meaningful and correct manner. This could involve, for example, sending a notification to the user or logging the event for later review. Similarly, errors within a queue listener might require special handling to prevent message loss or duplication. Implementing robust logging and retry mechanisms, as well as appropriate exception handling within both the transactional and queued contexts, is vital for building reliable, maintainable systems that smoothly handle the unexpected. Thorough testing, simulating various failure scenarios, is essential before deployment to ensure resilience and integrity of the system as a whole.

The choice between using transactions or queues often depends on the specific requirements of your application. Transactions are ideal for situations requiring strict data consistency and atomicity, such as financial transactions. Queues, however, are better suited for asynchronous operations where immediate response isn't necessary and handling failures gracefully and correctly is crucial, as in a system managing background tasks or large amounts of requests asynchronously. The use of transactions often leads to better performance in situations requiring immediate consistency; queues, on the other hand, generally offer better performance for high volume systems processing tasks independently. Understanding these trade-offs is crucial for making informed decisions during the design phase of your application. Often, the ideal solution may involve integrating both transactions and queues in a single system, utilizing the strengths of each approach.

Beyond fundamental implementation details, careful consideration should be given to the design of your components. If you're using transactions, ensure that the scope of the transaction aligns perfectly with the logical units of work. Too narrow a scope might lead to unnecessary overhead, while too broad a scope increases the likelihood of failures. Furthermore, avoid long-running transactions to minimize resource contention and enhance responsiveness. For queues, careful design of the message format is paramount. A well-structured message reduces ambiguity and enhances the maintainability of your listener. Effective monitoring of both transactions and queues is also crucial. Implementing monitoring tools allows you to identify potential bottlenecks and issues proactively, enabling timely interventions and resolution of issues before they significantly impact your application. Consider carefully how to best monitor your system, in terms of queue lengths, transaction times, and error rates.

5.2 Component Services

Component Services offer a powerful infrastructure for managing and deploying COM+ components. They provide a centralized location for configuring, monitoring, and controlling the lifecycle of your components. Understanding these services is essential for building robust and scalable applications. This involves not just deploying components, but also configuring their properties, managing their transactions, and handling potential failures.

Effective use of Component Services begins with understanding the Component Services administrative tool. This console allows you to browse the available components, view their properties, and make crucial configuration changes. For example, you can specify the identity under which a component runs, impacting its security context and access to resources. Careful consideration of this setting is crucial; incorrect configuration can lead to permission errors and application instability. Furthermore, the tool allows for detailed monitoring, enabling the identification of performance bottlenecks and the proactive resolution of issues. Proactive monitoring of your components can prevent unexpected failures and improve application reliability. Regular review of the logs helps you maintain your applications efficiently.

Beyond basic configuration, Component Services enables fine-grained control over component behavior. This includes the ability to configure transaction settings. Transactions ensure data integrity, particularly in scenarios involving multiple components or database interactions. Understanding different transaction types (e.g., local, distributed) and their implications is vital. Misconfigured transactions can lead to data inconsistencies or application crashes. The correct choice significantly affects the robustness and reliability of your application. You can leverage Component Services to maintain a central view of your entire system, providing for easier management.

Furthermore, Component Services facilitates the implementation of sophisticated queuing mechanisms. Message queues enable asynchronous communication between components, enhancing application responsiveness and scalability. This feature is especially useful in handling long-running operations or situations where components need to communicate without blocking each other. Properly implementing this requires careful consideration of queuing technology choices, message formats, and error handling strategies, significantly improving efficiency and avoiding deadlocks.

Deployment of COM+ components is streamlined through Component Services. The console provides tools for installing, uninstalling, and managing component versions. This ensures consistent deployment across different environments (e.g., development, testing, production). You can easily create packages for deployment and manage your application configurations. Furthermore, Component Services allows for efficient management of component dependencies and versioning, ensuring compatibility and stability across different versions.

Troubleshooting component failures often involves analyzing logs and events within Component Services. The detailed logging functionality provides invaluable information for diagnosing and resolving problems. This is essential for effective debugging; error messages and logs contain information for understanding failures, helping you prevent repetition and improve your software. They're an irreplaceable resource for understanding your system and maintaining it.

Finally, remember that proper security configuration is paramount when working with Component Services. Incorrectly configured security settings can expose your components to vulnerabilities. This means configuring appropriate access permissions and carefully controlling which users and processes can interact with your components. Always keep this paramount for your security and maintainability. Effective implementation safeguards data integrity and system stability. Regular security audits and updates are crucial in maintaining secure software deployment. Neglecting security can compromise the entire system.

In conclusion, mastering Component Services is fundamental to building robust, scalable, and maintainable COM+ applications. From initial configuration and monitoring to advanced queuing and deployment strategies, the tools and functionalities offered provide a comprehensive solution for managing the entire lifecycle of your components. By understanding and leveraging these features, you'll improve the development process, reducing errors and downtime. Furthermore, you'll increase your system's resilience and overall efficiency, essential for robust software architecture.

5.3 Deployment Strategies

Deploying COM components effectively requires a strategic approach encompassing several key considerations. Successful deployment isn't simply about copying files; it's about ensuring your components integrate seamlessly into the target environment, maintaining stability and performance.

Careful planning and execution are crucial for a smooth rollout and sustained operational success. Overlooking these steps can lead to frustrating debugging sessions and deployment failures, costing valuable time and resources.

Careful consideration of the target environment is paramount. The operating system, its version, and any specific system requirements must be meticulously examined. Dependencies, such as specific DLLs or other components, need to be accounted for; neglecting this can result in runtime errors that are often difficult to track down. Thorough testing on a system mirroring your deployment environment is a crucial step many developers overlook, a mistake that can lead to unexpected runtime issues when your components are put to work in their operational environments. Remember, thoroughness is your best friend during this phase.

Registration is the process of making your component known to the operating system. This is usually done using regsvr32.exe, but for larger deployments, automated registration tools or scripting should be considered for efficiency and repeatability. Manual registration is not suitable for large-scale deployments, and automation tools and scripts allow for a smooth, repeatable process, saving countless hours of manual intervention and minimizing human error. This step ensures that the system can locate and utilize the component as intended.

The choice of deployment method depends heavily on the application's scale and complexity. For simple scenarios, a straightforward manual installation might suffice. However, as complexity grows, more robust solutions are required. This could involve installers created with tools such as InstallShield or WiX, capable of handling dependencies, registry entries, and other aspects essential for smooth deployments. For more sophisticated scenarios, automated deployment systems, perhaps using scripting or configuration management software, are essential to ensure consistent, reliable installations across multiple machines.

Consider the security implications of your deployment. Is your component accessing sensitive resources? If so, access control lists (ACLs) and security considerations must be integrated into your deployment strategy. Without sufficient consideration for security aspects, potential vulnerabilities can be introduced, potentially exposing sensitive data or allowing malicious activities. This involves careful configuration of permissions and potentially code signing to protect your component and the system it's deployed on.

Post-deployment monitoring is frequently neglected yet incredibly important. Tracking component performance and stability after release is essential. Tools for monitoring performance metrics, logging errors, and gathering usage data can provide valuable insights into your component's behavior in a real-world setting. This data allows you to identify areas for improvement and address any issues arising after deployment. Proactive monitoring helps to ensure ongoing stability and performance. Ignoring this crucial step can leave you unaware of growing problems until they become critical.

Finally, comprehensive documentation is vital, both for your own reference and for anyone else who needs to work with your component. This includes clear installation instructions, explanations of dependencies, and detailed information about configuration options. Thorough documentation makes it significantly easier to install, maintain, and troubleshoot problems, regardless of whether you or others are involved in the process. Without it, troubleshooting will become a lengthy, unpredictable, and extremely frustrating experience.

Remember, a well-defined deployment strategy is as crucial as the component's design itself. Investing time and attention in this stage ensures a smooth rollout and fosters the long-term success of your COM components. Proper planning and execution significantly reduce deployment issues and contribute to the overall efficiency and stability of the entire software system. By understanding and implementing these strategies, you can build and deploy COM components with confidence. The payoff is a more reliable, stable, and easier-to-maintain system, minimizing time wasted on troubleshooting and maximizing your focus on the core functionality of your components. Don't underestimate the power of a well-planned deployment, it is a cornerstone of successful software engineering. This will ultimately save you time and effort in the long run.

6 Interoperability with Other Technologies

Successfully integrating COM components into diverse technological landscapes demands a deep understanding of interoperability. This chapter explores practical strategies for achieving seamless communication between your COM components and other technologies, enhancing the versatility and reach of your software. We'll delve into specific techniques, avoiding generalized advice, to ensure your success.

One crucial area is interoperability with the .NET framework. This involves bridging the gap between managed (.NET) and unmanaged (COM) code. The primary mechanism is through COM Interop, a feature built into .NET that allows managed code to seamlessly access COM components. You might use the Type.GetTypeFromProgID method to obtain a type from a COM object's ProgID, allowing instantiation and interaction with existing COM functionality within your .NET application. Careful consideration of data type mapping between managed and unmanaged environments is vital to prevent errors and ensure data integrity during interop calls. Remembering to handle potential exceptions thrown across this boundary is also critical for robust applications. Don't assume smooth sailing – comprehensive error handling is always a necessity.

ActiveX controls offer another significant avenue for interoperability. These controls, essentially COM components designed for visual integration into applications, can significantly enhance user interface functionality. Integrating ActiveX controls into your applications, whether they are Windows Forms, WPF, or web-based applications, leverages their pre-built functionality. You will discover that embedding them is relatively straightforward, using design-time interfaces or programmatic embedding. However, you need to carefully consider version compatibility and the potential security implications of using third-party ActiveX controls. Always prioritize controls from reputable sources and verify their security certifications. Thorough testing is also mandatory to prevent unexpected behavior or vulnerabilities.

Beyond .NET and ActiveX, successful interoperability extends to a broader range of APIs. This can involve interacting with databases, operating system services, or even third-party libraries. The approach often depends on the specific API. Some APIs may offer COM interfaces directly, facilitating seamless integration with COM components. In other cases, you may need to utilize bridging techniques, such as wrapper libraries, to create an interface that your COM components can interact with comfortably. Understanding the API's documentation and its communication protocols is paramount. Effective communication often requires careful mapping of data structures and proper error handling. Remember that inconsistent or missing documentation can significantly increase the complexity of this process.

For instance, consider integrating with a legacy system using a proprietary API. You might write a COM wrapper around this API, exposing a standardized COM interface to other parts of your system. This abstraction hides the complexities of the legacy API and keeps your main application logic cleaner and more maintainable. This modular approach increases the resilience of your system against changes in the underlying legacy systems. Always consider the trade-offs between tight coupling and loose coupling when integrating with any external API.

Effective interoperability also necessitates careful attention to data marshaling. This refers to the process of converting data from one format to another to enable communication between different processes or systems. Efficient marshaling minimizes overhead and improves performance. Understanding the different marshaling types, such as by value or by reference, is essential. Improper marshaling can lead to unexpected data corruption or performance bottlenecks. Consider if your system needs custom marshaling, which offers greater control but introduces increased complexity.

When dealing with potentially complex interactions, clear documentation and a well-structured design are absolutely crucial. Thorough testing, involving both unit and integration testing, is essential to ensure that the integration works as intended in different scenarios. This helps to identify and address any issues early in the development cycle. Use a version control system to track changes and ensure collaborative efforts are smooth. Never underestimate the value of thorough testing; it directly impacts the reliability and robustness of your integration. Remember that even minor issues during integration can lead to unpredictable system behavior. A methodical and documented integration approach is the key to avoiding such pitfalls. This also makes your work easier to understand and maintain over time.

6.1 COM and .NET Interop

Working with COM components from .NET applications presents a unique set of challenges and opportunities. The core of the challenge lies in bridging the gap between the unmanaged world of COM and the managed environment of the .NET Framework. This requires a deep understanding of how COM operates, its reliance on interfaces, and how .NET interacts with these interfaces. Understanding these nuances is crucial for successful integration. This chapter will provide practical guidance and techniques for effectively interfacing these two powerful technologies.

The most common approach to interacting with COM components from .NET involves using the Interop services built into the .NET framework. These services act as a bridge, translating calls from your managed code to the unmanaged code of the COM component. This translation process handles the complexities of marshaling data between these different memory models, freeing the developer from the low-level details. However, understanding the underlying mechanism is helpful for debugging and performance optimization.

Successful COM interop often begins with obtaining a reference to the COM component. This usually involves using the Type.GetTypeFromProgID() method, passing the Program ID (ProgID) of the COM component as an argument. The ProgID acts as a unique identifier for the component within the system's registry. This method returns a Type object representing the COM component, which can then be used to create an instance of the component using the Activator.CreateInstance() method. A simple example could be interacting with a hypothetical spreadsheet application via COM, enabling programmatic manipulation of its data.

However, simply obtaining a reference is only the first step. Working with the component's methods and properties requires careful consideration of the component's interface. Remember that COM relies heavily on interfaces; you will interact with the COM component through its defined interfaces. .NET's Interop services help to map these interfaces to corresponding .NET objects. This process is often automatic, but in more complex scenarios, you might need to explicitly define these mappings. For instance, if a COM component exposes an interface requiring a specific data type not directly mapped to a .NET type, careful consideration is needed.

Once you've obtained an instance of the COM component, interacting with it is similar to using any other .NET object. You call the methods and access the properties as defined by the component's interface. However, remember the importance of error handling. COM components might throw exceptions that need to be appropriately caught and handled within your .NET code. This robust error handling prevents unexpected crashes or data corruption within the application. Additionally, consider the lifetime management of the COM component. Ensure proper disposal using the IDisposable interface, especially relevant for components that manage external resources.

Advanced scenarios might necessitate custom marshaling. This involves specifying how data is passed between the .NET and COM worlds. This is particularly useful for complex data types or when you need to optimize performance. Custom marshaling is more complex to implement than the default marshalling. It may include creating custom marshalers using the IMarshal interface or employing other techniques to tailor data exchange precisely. Custom marshaling significantly improves performance and addresses specific data type concerns in COM to .NET interoperability. However, it requires a greater understanding of COM's internal workings.

Finally, security should always be a top concern when working with external components. COM components may have security implications, particularly if they interact with system resources or sensitive data. It's crucial to validate the source and integrity of any COM component you integrate. Proper security measures and validation are critical components in ensuring the safety and stability of the application. Understanding these security concerns is paramount for developers, and appropriate practices should be employed to mitigate any security risks. This might involve using code access security measures or implementing careful input validation within the application.

In conclusion, integrating COM components into your .NET applications offers powerful possibilities for extending functionality and leveraging existing codebases. However, careful planning and a thorough understanding of COM principles and .NET's Interop services are crucial for success. Remember that mastering this area involves understanding the nuances of interface definitions, marshalling, error handling, and security considerations, all within the context of managing resources effectively. Mastering these elements will unlock the full potential of COM interoperability in your .NET projects.

6.2 Working with ActiveX Controls

ActiveX controls, a cornerstone of COM, offer a powerful mechanism for extending applications with reusable components. They provide a straightforward way to integrate pre-built functionalities into your projects, enhancing user experience and speeding up development. Understanding their intricacies is key to harnessing their full potential, thereby avoiding common pitfalls that can hinder development efforts.

Creating a robust ActiveX control necessitates a meticulous design process. This starts with defining a clear purpose and functionality. What specific tasks will your control perform? How will it integrate seamlessly with the host application? These questions, answered before a single line of code is written, contribute significantly to the control's overall success, ensuring efficient resource management and a streamlined workflow. Carefully planning the interface, deciding on the properties and methods, and foreseeing potential interactions with the host application's environment are fundamental steps in developing a versatile and adaptable control.

The implementation phase demands a deep understanding of COM's underlying principles. Proper event handling is crucial for enabling dynamic interaction between the control and the host application. Consider using well-established patterns, like the observer pattern, to manage events effectively. This ensures efficient communication and prevents unforeseen complications. The control's responsiveness and stability depend heavily on how events are managed; thus, proper event handling should not be overlooked. Remember to handle errors gracefully, providing informative feedback to the user and preventing abrupt crashes. Thorough error handling is an absolute necessity for any robust control.

Testing is an essential part of the development lifecycle. Testing your ActiveX control rigorously in different environments and scenarios helps to identify potential bugs and compatibility issues early. A robust testing strategy will uncover issues related to memory management, event handling and stability under different loads. Use a variety of host applications to ensure that the control functions correctly across different platforms. This will ensure broader usability and wider compatibility. A well-tested control provides higher confidence, therefore a substantial improvement in the overall quality of the product.

Deployment of your ActiveX control also requires careful planning. Consider using industry standard packaging and deployment technologies. Understanding the security implications is vital, ensuring users can install and use the control safely without compromising their system's security. Ensuring the control's installation process is smooth and intuitive is crucial for positive user adoption. A smooth and clear installation process is critical for user experience and ultimately, the control's widespread acceptance.

Integrating ActiveX controls into your applications often involves working with various development environments and languages. Understanding how to interact with the control's properties and methods from different languages, such as C++, VB.NET, or JavaScript, is important for maximizing its versatility. Each language has unique mechanisms for interfacing with COM objects, thus the correct techniques and syntax need to be carefully implemented to achieve seamless integration. This cross-language compatibility is essential for broadening the scope of your application's use. By adapting your control to different programming environments, the number of potential uses increases significantly.

Beyond the basics, optimizing the control for performance is a critical step. Efficient memory management prevents memory leaks and enhances the responsiveness of the application. Employ techniques to minimize unnecessary resource consumption. Analyzing the control's performance in real-world scenarios using profiling tools, can reveal bottlenecks, leading to targeted optimization strategies. Performance considerations should be integrated into the design process, to ensure that the control is as efficient as possible, ultimately improving the overall application performance.

Furthermore, consider security implications throughout the development lifecycle. Secure coding practices are essential to prevent vulnerabilities. Regular security audits, performed by experienced professionals, are a proactive measure to help reduce the risks associated with security vulnerabilities. By employing various strategies such as input validation and secure data handling, the robustness of your control and the security of your applications can be enhanced. Remember, a secure control is a fundamental aspect of a trusted application. Security should always be a priority.

Finally, remember to meticulously document your ActiveX control. Clear and concise documentation is vital for maintainability and ease of use. Detailed explanations of the control's functionalities, usage instructions, and any specific considerations are critical for wider adoption. Well-written documentation enables other developers to easily use and extend your control, thereby fostering collaboration and community engagement. Moreover, comprehensive documentation makes maintenance much more straightforward for future revisions and updates. This documentation should be readily accessible and easy to understand.

6.3 Integration with Other APIs

Integrating your COM components with other APIs opens up a world of possibilities, extending their functionality far beyond what's achievable in isolation. This isn't simply about connecting; it's about seamless, efficient collaboration. Consider, for instance, integrating with a third-party payment gateway. This single integration could transform your application from a simple utility to a fully functional e-commerce platform. The key lies in understanding the target API's structure and communication protocols, which often involve careful consideration of data formats, and a smooth handling of asynchronous operations. This integration strategy is critical for building a robust and flexible system.

Successfully bridging the gap between your meticulously crafted COM component and an external API usually hinges on meticulous planning. You'll need a clear understanding of both your component's capabilities and the API's limitations. Thoroughly review the API's documentation; this often-overlooked step is the foundation of a successful integration. Pay attention to authentication methods, rate limits, and any specific requirements for data formatting. Ignoring these details can lead to frustrating debugging sessions and integration failures. Prioritize thorough research and testing throughout the process.

Different APIs employ various communication protocols. REST, often using JSON or XML for data exchange, is extremely prevalent. Others may rely on SOAP, gRPC, or even proprietary protocols. Your integration strategy must align perfectly with the chosen protocol. For instance, if the API uses REST, you'll likely use HTTP requests within your COM component to interact. This necessitates careful handling of HTTP responses, including error codes and the parsing of the returned data. Remember that effective error handling is paramount, regardless of the protocol employed. A robust system should gracefully handle various error scenarios, informing users appropriately and preventing application crashes.

Let's say your component needs to interact with a database. Instead of directly embedding database-specific code within your component, consider leveraging a database API or an ORM (Object-Relational Mapper). This approach promotes better separation of concerns and makes your component more adaptable. You can swap databases without modifying the core component code, simply by altering the API interaction layer. This flexible architecture significantly improves maintainability and allows for easier future upgrades. Prioritizing modular design will ensure your component remains robust and adaptable to changing requirements.

Security is always a critical aspect, and it's especially true when integrating with external APIs. Never transmit sensitive data like passwords or credit card numbers in plain text. Always use secure protocols such as HTTPS and robust encryption techniques. Implement proper authentication and authorization mechanisms, verifying the API's responses and the identity of the client. Regular security audits are crucial to identify and address potential vulnerabilities, protecting your application and user data. Proactive security measures are an essential aspect of responsible software development.

Testing your integration thoroughly is vital to prevent issues in production. Begin with unit tests to verify individual components of your integration code. Progress to integration tests which examine the interaction between your component and the API. Finally, conduct thorough end-to-end tests, simulating real-world scenarios and encompassing the entire application workflow. Employ techniques such as mocking to isolate specific parts of the system, making debugging and testing more manageable. Remember, rigorous testing saves time and prevents costly errors later.

Effective documentation is as vital as the integration process itself. This documentation should outline the integration strategy, including API credentials, communication protocols, and error handling mechanisms. Detailed API interaction sequences should be clearly explained, along with the handling of various response codes. This documentation will not only help you troubleshoot problems but also facilitate collaboration and knowledge sharing among team members. Well-structured documentation is an integral part of the software development lifecycle. Consider the broader context of your application and how integration influences the overall user experience. The speed and reliability of your API integration are directly related to your application's overall performance and user satisfaction.

7 Advanced Debugging and Troubleshooting

Debugging COM components can be challenging, demanding a methodical approach. Start by replicating the error consistently; this ensures your debugging efforts target the root cause, not a fleeting anomaly. Carefully examine error messages, looking for clues like specific HRESULT codes. These codes often point directly to the source of the problem, providing a starting point for investigation. Don't overlook the importance of logging; detailed logs, including timestamps and relevant contextual information, can be invaluable for pinpointing the exact moment a problem arises. Consider using a debugger that supports COM, allowing you to step through component interactions, inspect variable values, and identify problematic code sections. Efficient debugging necessitates a combination of careful observation, systematic analysis, and the right tools.

Advanced debugging often requires examining the component's interactions with other parts of the system. Use a tool like Process Monitor to observe file system accesses, registry modifications, and network activity; this provides a comprehensive overview of the component's behavior. Pay close attention to timing issues, as race conditions or deadlocks can manifest as unpredictable errors. Careful analysis of the component's thread usage is crucial. Understanding how the component interacts with different threads can illuminate synchronization problems. Consider using tools that allow you to visualize thread execution, helping to identify points of contention and possible deadlocks. If working with a multi-threaded environment, you might need advanced debugging features like thread breakpoints and stack trace analysis. Remember, a systematic approach is key.

Troubleshooting common issues frequently involves understanding the COM infrastructure. Memory leaks, for example, often manifest as slow performance or crashes. Use tools such as the Windows Performance Analyzer to monitor memory usage, identifying areas of excessive memory allocation or insufficient deallocation. Pay attention to resource handles; failure to release resources correctly can lead to resource exhaustion and system instability. COM object lifecycle management is crucial. Improper creation or release of COM objects, like failing to release IUnknown correctly, can cause major problems. Double-check your code for proper reference counting and ensure your object lifecycle adheres to COM guidelines. Dealing with marshaling errors can be tricky, but many problems stem from incorrect data types or interface mismatches. Verify that the data types being marshaled are compatible between the client and server components. Use the debugger to meticulously examine the data being passed and its transformation during the marshaling process. Consider using custom marshaling if standard marshaling proves insufficient. Custom marshaling provides more fine-grained control, allowing you to handle complex data structures effectively. Thoroughly document your marshaling logic and any custom marshaling implementations for future reference and maintenance purposes. This helps in troubleshooting potential problems.

When troubleshooting, consider external factors impacting your COM components. Are there any environmental variables affecting component behavior? Are there conflicts with other software or drivers? Checking system events, looking for warnings or errors related to your component or its dependencies, can provide essential clues. If the issue seems related to a specific configuration, meticulously compare the problematic setup with a working environment; this helps pinpoint the source of the inconsistency. Consult official documentation and online resources; many solutions to common COM problems can be found online. Remember, the community is a great asset in these situations.

Addressing security vulnerabilities requires a proactive approach. Ensure your components use appropriate authentication and authorization mechanisms, preventing unauthorized access. Verify that data is properly validated and sanitized, preventing injection attacks. Use code analysis tools to identify potential security vulnerabilities and adopt secure coding practices. Always thoroughly test your component's security, simulating various attack scenarios to assess its resilience. Regular security updates and patching are essential to protect your components against known exploits. A secure component design requires a layered security strategy throughout development and deployment.

Finally, thorough testing is critical. Write unit tests to verify individual component functionalities, integration tests to check the interplay between components, and system tests to evaluate the overall behavior within the complete system. Test under various conditions, including heavy load and stress scenarios, to uncover hidden issues. Use test frameworks designed for COM components. Effective testing prevents potential issues from surfacing in production, ensuring a higher level of stability and reliability. Remember, a well-tested component significantly reduces troubleshooting efforts in the future. Don't underestimate the importance of creating a solid test plan; planning ensures comprehensive coverage of all scenarios.

7.1 Debugging COM Components

Debugging COM components can be challenging, but a systematic approach significantly improves your chances of success. Start by reproducing the problem consistently. This often involves creating a simple, repeatable test case that isolates the issue. Don't jump to conclusions; meticulously document every step, including the exact sequence of events leading to the error. This detailed record proves invaluable for later analysis and identifying patterns. Remember, the simplest reproduction case will often provide the clearest view of the root cause.

Once you have a repeatable test case, leverage the power of the debugger. Familiarize yourself with its features, especially breakpoints, watch expressions, and call stack analysis. Set breakpoints in key areas of your code, like the component's initialization, method entry points, and resource release. Step through your code line by line, examining variable values and the flow of execution. Pay close attention to how data is passed between components; often, marshaling issues are the culprit. If using a sophisticated IDE with debugging capabilities for COM components, be sure to thoroughly research and learn its specific tools and techniques. The debugger should reveal anomalies not obvious in the code itself.

The debugging process itself is a cycle of observation, hypothesis, and testing. You observe the behavior of your component in the debugger, formulate a hypothesis about the cause of the error, and then test this hypothesis by making changes to your code or its environment and observing the impact. This iterative refinement helps eliminate possibilities until you find the issue. Don't neglect logging. Strategically placed logging statements provide a valuable window into the internal state of your component. Record key events, data values, and function calls at critical points. When errors occur, this log offers crucial insights, often revealing where and why things went wrong. Comprehensive logging proves beneficial, not only during debugging but also throughout the component's lifecycle.

Next, investigate memory leaks. COM components rely heavily on proper resource management. Use memory leak detection tools to identify any unfreed memory. Memory leaks manifest in several subtle ways; they may not immediately cause crashes, but they can lead to performance degradation over time, making it difficult to pinpoint the problem's source. Tools like those included with your debugger or those available as independent utilities, such as those built-in to the Visual Studio IDE, can greatly speed up this process. These tools have different capabilities, so it is worth exploring a few before committing to one or another. It's crucial to address memory leaks as they can dramatically compromise the stability and responsiveness of your applications.

Another common cause of issues is improper handling of interfaces. Ensure that your component correctly implements all interfaces and adheres to the specification. Verify that all references to interfaces are properly released using Release() or equivalent methods. Failing to release references can cause circular references, leading to memory leaks or unexpected behavior. The documentation of your framework will provide a good starting point for identifying these types of issues. This often involves a deep understanding of reference counting mechanisms.

If your component interacts with other components, consider the potential for miscommunication. Check the versions and compatibility of all involved components. Make sure that data is being marshaled correctly between processes or threads. A subtle incompatibility or a mismatch in data types can often lead to seemingly random errors, and thus it's best to ensure that the versioning of components aligns and all interactions are consistent and free from unexpected changes. This can be a significant source of errors for even

the most experienced developers.
Finally, don't overlook the importance of testing. Unit tests ensure that your individual methods function as expected in isolation. Integration tests examine how your component interacts with other parts of your application, and System tests confirm the overall functionality and stability of your system. This multi-layered approach significantly enhances your capacity to detect errors and to ensure that your component is reliable and robust across various scenarios and usage patterns. Remember, testing is an ongoing process, not a one-time effort. Thorough testing improves overall quality and makes it easier to locate potential defects. Through this methodical approach to debugging, you'll significantly improve the stability and reliability of your COM components.

7.2 Troubleshooting Common Issues

Many COM component issues stem from seemingly minor coding oversights or misconfigurations. A common problem involves incorrect interface definition, leading to type mismatches or unexpected behavior during runtime. Double-check your IDL file meticulously; even a single misplaced semicolon can cause cascading errors. Remember that consistency is key; adhere rigorously to naming conventions and data type definitions across all interfaces to avoid confusion and ensure seamless interaction between components. Consider using a static code analysis tool to catch potential errors before they escalate into serious runtime problems. Failing to properly manage memory is another frequent pitfall. Ensure every CoCreateInstance call is matched with a corresponding Release. Memory leaks, while sometimes subtle, can cripple performance over time and lead to unpredictable crashes. Implement rigorous memory management practices; adopting a disciplined approach of resource acquisition is initialization (RAII) greatly reduces the chances of such errors. Thorough testing and profiling are essential to identify and address memory issues early in the development cycle.

Class factory registration errors are another common source of frustration. If your component isn't registering correctly, it simply won't be available to other applications. Verify that your class factory is correctly implemented and registered in the registry using the appropriate methods, making certain your component's CLSID and ProgID are accurately defined. Ensure your registration process accounts for potential conflicts with pre-existing entries, especially when working within a shared environment or deploying updates. Employ automated testing to validate successful registration, which can reduce the manual labor and human error involved.

Marshaling problems often manifest as access violations or inexplicable failures during inter-process communication. Incorrectly marshaled data can lead to corrupted data structures and program instability. Carefully review the marshaling attributes defined within your interfaces, ensuring they accurately reflect the data types being exchanged and employing appropriate marshalling techniques depending on the complexity of your data. Rigorous testing with various data sets helps reveal hidden marshalling vulnerabilities. Thorough logging during marshaling can pinpoint the precise point of failure.

Thread safety issues are particularly challenging to diagnose, frequently resulting in race conditions and data corruption. Ensure your component is correctly designed to handle concurrent access from multiple threads, potentially by implementing appropriate synchronization mechanisms such as mutexes or critical sections. You should also ensure the consistency of your thread models across all components. When dealing with multithreaded scenarios, consistently employing these critical safeguards is paramount. The use of a debugging tool that allows thread-level debugging capabilities becomes indispensable in tackling these situations.

Another frequent cause for concern is improper error handling. Failure to correctly handle exceptions can lead to the program abruptly terminating, leaving users with an unexplained error. Thoroughly implement robust error handling mechanisms within your components. Implement comprehensive logging strategies to record errors and exceptions for later analysis. These practices contribute to creating more robust applications. Testing under diverse conditions allows for the identification of potential issues, improving the reliability of your application.

Finally, deployment complexities can often cause unexpected problems, particularly when dealing with dependencies between components. Carefully plan your deployment strategy, paying close attention to the versioning of your components and any dependencies your program might have. Use of a deployment tool to streamline and automate the deployment process can greatly enhance efficiency and reduce errors that often stem from manual processes. It's prudent to follow established practices for creating version-compatible components and thoroughly test deployment under various scenarios to ensure smooth installation across different environments.

www.ingramcontent.com/pod-product-compliance
Lightning Source LLC
LaVergne TN
LVHW022023200926
848481LV00007B/2506
* 9 7 9 8 3 4 8 5 5 7 8 0 5 *